U.S. ARMY DRILL SERGEANT HANDBOOK

January 2009

Foreword

The Center for Army Lessons Learned has partnered with the Basic Combat Training Center of Excellence and Fort Jackson to assist drill sergeants to prepare, train, and lead our new Soldiers in their transition to ground combatant.

Mistakes occur more frequently during the first and last cycles of your drill sergeant tour. During the first cycle, you are one of the least experienced drill sergeants and lack knowledge of the command climate. This handbook will alert you to some common errors so you do not make them. These mistakes can be detrimental to your Army career and negatively affect the lives of your Soldiers. During your last cycle, you may get too relaxed and have a false sense of security. Do not fall into that trap. We, as leaders, must give you all the tools you need for success.

This handbook is designed to help you through one of the most difficult and challenging missions of your Army career—teaching and molding our future Soldiers to train, fight, and win in the Global War on Terrorism. It gives "street smart" advice from more experienced current and former drill sergeants who also want you to succeed. In this handbook, they show you how to be a better leader, handle the many pressures of your position, and overcome numerous training obstacles.

Your excellence as a drill sergeant equates to hundreds of Soldiers' lives saved. Being a drill sergeant is a serious and tough job, but with perseverance and dedication, you can do it!

BRADLEY W. MAY
Brigadier General, U.S. Army
Commanding

Drill Sergeant Handbook

Table of Contents

Introduction	1
Chapter 1. Knowledge	3
Chapter 2. Skills	19
Chapter 3. Abilities	29
Chapter 4. Attitude	37
Appendix A. References	41
Appendix B. Drill Sergeant Peer Evaluation Form	43
Appendix C. Soldier Assessment Form(IET Soldiers)	45
Appendix D. Basic Combat Training Smart Card	47
Appendix E. Major General John M. Schofield's Definition of Discipline	49

Center for Army Lessons Learned

Director	Colonel Robert W. Forrester
Project Analyst	Michael Pridgen
Production Coordinator	Valerie Tystad
Editor	Jenny Solon
Graphic Artist	Dan Neal
Printing Support Liaison	Candice Miller

The point of contact for this handbook is Mr. James E. Walthes, Director, Doctrine and Training Development Division (803-751-6511/1137 or DSN 734-6511/1137; jim.walthes@us.army.mil).

CENTER FOR ARMY LESSONS LEARNED

The Secretary of the Army has determined that the publication of this periodical is necessary in the transaction of the public business as required by law of the Department.

Unless otherwise stated, whenever the masculine or feminine gender is used, both are intended.

Note: Any publications (other than CALL publications) referenced in this product, such as ARs, FMs, and TMs, must be obtained through your pinpoint distribution system.

Introduction

Being a drill sergeant may be the most challenging and rewarding assignment a noncommissioned officer will ever experience during his military career.

While training initial entry Soldiers to fight and win in today's Global War on Terrorism, drill sergeants must embody and reflect the Army's values and standards. They must also be:

- Effective communicators.
- Sound administrators.
- Motivators.
- Proficient instructors.
- Effective coaches, mentors, and counselors.

This handbook is designed to help new drill sergeants conquer the many challenges of their assignment and succeed in their mission of training Soldiers.

Primary sources for this handbook are current drill sergeants from The United States Army Drill Sergeant School, Fort Jackson, SC, and former drill sergeants at the Basic Combat Training Center of Excellence. Special thanks to each installation Drill Sergeant of the Year for their valued input. A full list of references is located at Appendix A.

Chapter 1

Knowledge

On Orders for Drill Sergeant Duty?

Prepare yourself for your new mission:

- Learn the Drill Sergeant Creed and the Soldier's Creed.

- Get in shape (mentally as well as physically) and eat right while preparing yourself for a challenging but rewarding experience. Be open-minded.

- Prepare your Family for a challenging assignment. Explain that you will be required to work late nights, early mornings, weekends, and holidays. Connect your spouse with local Family resource centers.

- Talk to other current or former drill sergeants in your unit or area. They will give you helpful advice and support.

- Be familiar with the talk-through, step-by-step, and by-the-numbers methods of instruction located in the Module Handbook (at <http://www.jackson.army.mil/units/drill/index.html>) for the following:

    ◦ Stationary drills

    ◦ Facing movements at the halt

    ◦ Steps and marching

    ◦ Basic manual of arms

    ◦ Advanced manual of arms

    ◦ Squad drill

    ◦ Platoon drill

- Study Field Manual (FM) 3-22.20, *Army Physical Readiness Training*. Focus on executing the exercises.

- Be familiar with United States Army Training and Doctrine Command (TRADOC) Regulation (TR) 350-6, *Enlisted Initial Entry Training (IET) Policies and Administration*, and TR 350-16, *Drill Sergeant Program*.

- Have confidence in yourself. If you are an outstanding noncommissioned officer (NCO), then you will be an outstanding drill sergeant. If you are not an outstanding NCO, determine what you need to improve with a solid action plan.

- Learn a couple of marching and running cadences prior to attending Drill Sergeant School. Practice calling cadence with your current unit.

- Study the drill sergeant history to get a different perspective of the organization and a new level of pride. Visit the Basic Combat Training (BCT) homepage at <http://www.bct.army.mil> to see the history of the Drill Sergeant School or the United States Army Drill Sergeant School (USADSS) Website at <http://www.jackson.army.mil/units/drill/index.html> for more information.

When you report to your unit:

- Be motivated and open-minded. Drill sergeant duty is what you make of it. Do not be afraid to ask questions or say "I don't know."

- Look for great leadership. Seek a mentor with similar leadership styles and abilities. Pick his brain—find out how he does things so well.

- Find out what works for you. Surround yourself with positive, motivated leaders.

- Observe and then jump in with both feet while being critiqued by the more experienced drill sergeants. Seek feedback from your peers. Volunteer to explain or demonstrate so you can ultimately become the training expert.

- Keep the lines of communication open between you, your peers, and your chain of command.

- Keep your supervisor informed of any Family issues or appointments that can distract you from training Soldiers. Make sure your Family is settled. Ask your supervisor for more time up front if you need it.

- Stress the Army Values and never violate them! Enforce respect and self-discipline.

- Be knowledgeable on FM 3-22.20, TR 350-6 (especially Chapter 2, which discusses phase training and the associated goals and privileges for BCT, advanced individual training [AIT], and one station unit training [OSUT] Soldiers), and TR 350-16.

- Consult subject matter experts (SMEs) on tactics and become an SME yourself. Do not pretend to know what you are talking about.

- Seek advice from the more experienced drill sergeants on how and when to counsel your Soldiers.

- Check your emotions at the door—be thick skinned! Remember your own BCT experience (what frustrated you and what motivated you).

- Take what you learned at the USADSS and integrate it into your scope of training.

New Drill Sergeants

As a new drill sergeant, you should:

- Avoid being extra tough or mean in order to gain respect from the other drill sergeants. You will be judged on your knowledge and abilities.

- Speak up. If you have a question, ask! Go directly to the SME for assistance, advice, or guidance prior to instructing a class.

- Take control of the situation, and do not let yourself be intimidated.

- Avoid using gender to circumvent a duty or task.

- Design your own techniques on how to handle Soldiers. Be direct and fair to ensure all Soldiers know the rules and regulations. Let them know what will and will not be tolerated or accepted.

- Counsel all violations.

- Avoid excessive screaming; it becomes a distraction and could make your job difficult.

- Ensure your battle buddy supports your decisions; you are part of a team.

Outcomes-Based Training

Outcomes-based training (OBT) is the approach to training that stresses the desired end state/outcomes for all tasks; this approach equally develops mental intangibles, attributes, and a deeper understanding of the tasks being trained as they apply to combat, personal accountability, and Warrior Ethos. Every military task trained should be based on mastering fundamentals required for excellence and combat success, how the task ties into other supported or supporting tasks/functions, how the task applies to the individual's personal responsibilities, and how the execution of the task develops and validates the Warrior Ethos.

Every individual task that we train should be viewed holistically, combining all possible associated training benefits. This approach allows trainers the opportunity to train supplemental/related individual tasks simultaneously, to teach how these tasks work together to achieve superior synergistic results, and to tie into training the development of multiple intangible attributes which link directly to the Warrior Ethos. While individual OBT will need focused attention (specific movements, zeroing procedures, plotting coordinates, etc.), trainers should always tie back into the training session all related factors: how the task ties into others, the combat application (if applicable—some tasks have little combat application but are important for the soldierization process), personal responsibilities tied to tasks, and the Warrior Ethos.

> All of these should be linked to the specified individual tasks and explained in enough detail to provide the fundamental knowledge to allow Soldiers to understand fully, make future decisions as conditions change, and provide input for better methods. It is implied that throughout the execution of OBT, trainers must teach the fundamentals/principles to a level that ensures Soldiers retain the knowledge required to understand the task completely and gain personal accountability in the task execution.
>
> Trainers must consistently talk with their Soldiers to ensure that this deeper understanding is resonating with each individual and that they have explained in enough detail that the Soldier truly understands what performance is expected.
>
> They must measure how well the Soldiers understand through the Soldiers' demonstrated application, their ability to explain, the level of personal responsibility they take with task execution, and the level of excellence they achieve. As the trainer moves Soldiers through individual tasks, they must ensure the Soldier understands how tasks support additional or related tasks, building towards holistic achievements for multiple tasks. This approach will assist unit leaders as they develop training into collective requirements; Soldiers will better understand the linkage of individual tasks to collective tasks. They will also better understand how individual performance ties into unit success and mission accomplishment. When executed properly, this methodology will ultimately lead to task proficiency, development of intangible attributes, knowledge-based understanding, and an understanding of how all tasks combine for unit success.
>
> —Comments from a Command Sergeant Major

These concept slides may also help:

> **Working Definition for Outcomes-Based Training**
>
> Outcomes-based training is a philosophical approach to military training that stresses the end state of the Soldier's mental intangibles, attributes, and skills required by the commander for combat. The training is guided by the commander's intent and unit initiative to obtain the greatest effectiveness and not focused on process-driven requirements.

DRILL SERGEANT HANDBOOK

> **Working Principles of Outcomes-Based Training**
>
> Training designed to obtain desired combat-required outcomes in Soldiers to include mental intangibles, attributes, and skills.
>
> Develops effectiveness instead of fixating on efficiency and restrictive processes.
>
> Training allows Soldiers to make appropriate decisions and to understand the "why" behind various tasks.
>
> Commanders and cadre assess Soldier outcomes to ensure desired results are obtained.

Outcomes-based training is a tool commanders use to measure their Soldiers' and units' progress and make adjustments as they see fit. By carefully analyzing and evaluating training, commanders can determine if Soldiers are obtaining the desired outcome.

Units across United States Army Accessions Command (USAAC) have adopted outcomes-based training in different ways. The five elements depicted in Figure 1-1 have been approved by USAAC; however, they can be modified by units to fit their particular missions and desires.

Figure 1-1

Use outcomes-based training posters throughout your company or training area to inspire Soldiers.

Refer to the outcomes-based training concept before, during, and after training sessions. Believe in the concept, and you will come across to the Soldiers as genuine. Incorporate everything you do or say with the desired outcomes. Always reassess yourself so you can give the most effective training.

An additional week of basic combat training is provided for commanders to meet outcomes-based training requirements in FY09. No additional training is added to the program of instruction, allowing commanders to schedule training events based on the unit's training performance.

New Soldier Processing

As a drill sergeant, you should take the time to know and understand the entire Soldier processing flow—recruitment, military enlistment processing station (MEPS), Reception Battalion, BCT/OSUT/AIT, and reporting to the first duty station. You have the most important role in the development process:

- Develop a good rapport with the reception personnel.

- Correct **every** deficiency as early as possible (especially in the first two weeks). Make sure Soldiers understand why they are being corrected and what the action could cause. For example, new Soldiers who do not make their bunks to standard display a lack of attention to detail and time-management skills. Remember that every Soldier brings a certain amount of baggage with him that could possibly interfere with his performance during training.

- Ensure Soldiers know that they are now a part of the Army Family and that you care about their safety and well-being, are willing to help, and believe in them and their abilities.

- Take time to learn about new Soldiers. Do not assume they are weak or "worse" than Soldiers you trained in the past. Some Soldiers will be older than you; some will still be in high school

Reception Battalion Operations

Understand the Reception Battalion's role in processing new Soldiers and how the battalion can assist you.

Typical arrival schedule:

- First four hours:
    - Meet your drill sergeant
    - Initial processing
    - Amnesty room
    - Physical training (PT) uniform issue
    - Late dinner
    - Assigned to a company

- Day 1:
    - Wake up
    - Tuberculosis Tine test
    - Pay (via EZPay card)
    - Haircut
    - Medical processing (DNA, blood, optometry, audiology)
    - Post exchange (shoe type and size check)
- Day 2:
    - Personnel administration branch
    - Common access card
    - Clothing initial issue point (CIIP)
    - Dental processing
    - Top Secret security screening
- Day 3:
    - PPDE evaluation (medical)
    - Immunizations (mumps, measles, rubella; polio; meningitis; flu; tetanus)
    - Optometry
    - Photographs
    - PPDE brief (medical)
    - Correct various discrepancies
    - Moment of truth ceremony
    - Ship to BCT

Expedited/Blended Reception Model

Expedited/blended reception is the process of receiving, processing, and training Soldiers from the moment of arrival. The blending of reception and training immediately starts the transformation process from civilian to Soldier. BCT/OSUT drill sergeants command and control all newly arriving Soldiers from the moment they arrive on the installation. This process creates an environment where BCT/OSUT cadre set and enforce standards from the Soldier's night of arrival through graduation day. The one BCT/OSUT cadre has immediate ownership of the new Soldier.

BCT units receive all Soldiers arriving Monday-Thursday. Soldiers who arrive Friday-Sunday will be processed by the reception battalion and handed off on Monday to their BCT unit.

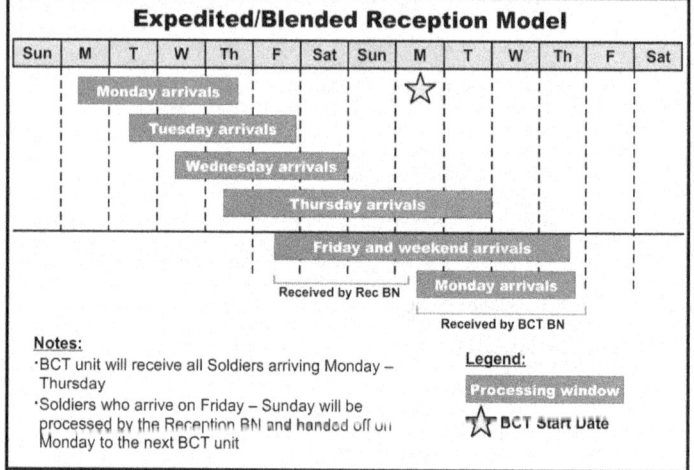

Figure 1-2

Use expedited/blended reception to get to know your Soldiers. Read the Soldiers' personnel data forms or talk to them about why they joined the Army, any Family issues they have, names of their children, and so on.

Develop an effective reception so Soldiers are focused on being Soldiers and not on the process. The goal of this program is for Soldiers to receive a best first impression of the Army and embrace their decision to join.

Expedited/blended reception eliminates multiple chains of command, establishes defined standards, and reduces a Soldier's confusion. It increases drill sergeant/Soldier fusion; you understand the Soldier and begin to teach, coach, and mentor faster. Expedited/blended reception reduces a new Soldier's emotional turbulence.

Conduct a one-on-one informal counseling session between you and the Soldier within 48 hours of the Soldier's arrival:

- Find out why your new Soldiers really joined the military.
- Conduct in-depth counseling to determine the success of the platoon for the next several months.

One of the biggest challenges in expedited/blended reception is communication between the Reception Battalion cadre/staff and the drill sergeants. Reception Battalion cadre/staff need to support the drill sergeants' needs/requests, and drill sergeants need to coordinate (as early as possible) with the Reception Battalion cadre/staff and be open to suggestions.

Some advantages to expedited/blended reception include the following:

- You are the primary trainer and have the ability to manage the reception process and intertwine key training tasks associated with BCT while capturing unused time to instruct Soldiers on vital warrior tasks.
- Drill sergeants can quickly begin training Red Phase tasks (immersion in the Army's core values, traditions, Warrior Ethos, ethics, individual BCT skills, teamwork, and PT) earlier.
- Time spent at the Reception Battalion is reduced or eliminated.
- Drill sergeant is upfront during new Soldier processing so that problems (clothing issue, medical problems, pay, boots, glasses, etc.) are corrected at the earliest possible time.
- Soldiers have no time to pick up bad habits or advice from others at the Reception Battalion.
- Soldiers experience one BCT standard. Soldiers are more responsive when they have one chain of command leading them from start to finish.
- Soldiers arrive at reception as close to Army Training Requirements and Resources System (ATRRS) start date as possible.
- Transportation requirements for moving new Soldiers are reduced. The ultimate goal is movement from arrival by airplane/bus to BCT barracks. Reception barracks may be necessary based on installation needs.
- Expedited/blended reception could possibly limit your cycle breaks. For example, you could graduate a class on Friday, ship them on Saturday, and then pick up a new class on Monday (depending on the time of year).

Drill and Ceremonies

Use cadence to reinforce learning while marching to and from all training sites or during PT (dress-right-dress, cover, recover, right face, left face, etc.).

Teach Soldiers as much drill and ceremonies as possible; go beyond the minimum requirements; take advantage of breaks in training. Help Soldiers improve through continuous practice or drill and ceremonies competitions. Your training will pay off when you proudly march them at graduation.

There are many advantages to using your Junior Reserve Officer Training Corps (JROTC) Soldiers to demonstrate drill and ceremonies:

- They know most of the techniques and can show other Soldiers how to perform them correctly.

- They provide a good visualization, so you will have less to correct.

- They show how receptive the Soldiers are; they learn faster using this approach, and it gives them an opportunity to develop their leadership skills.

Army Knowledge Online (AKO) Website

During registration at the MEPS, Soldiers are assigned an AKO account and receive specific instructions on its use. During training, Soldiers have little opportunity to use their AKO accounts. They normally forget their AKO login information by the time they get to AIT. Soldiers are also not familiar with the "MyPay" system. You should brief them on both AKO and MyPay prior to completing training.

Recognizing the Soldier's Potential

Techniques for using Soldiers to assist training include the following:

- Identify all Soldiers early so you can use them in different leadership positions (platoon leaders, squad leaders, or team leaders). Pinpoint their leadership traits/strengths and leverage as much as possible.

- Have the platoon come up with its own motto; it gives them a sense of ownership and instills pride.

- Have Soldiers assist you in preparing packets for the next cycle; it makes for a smoother reception.

- Have Soldiers write a "real deal" history of themselves (stress how important it is for them to tell the truth no matter how difficult it may be). This history will assist you in identifying their strengths and weaknesses as well as how to individually deal with and assist them.

- While awaiting transportation or during a break after training, brief Soldiers on upcoming training events so they can start preparing for the next day. Encourage Soldiers to help each other.

- Use platoon, squad, and team leaders to help make your job easier and to prepare them for higher leadership positions.

DRILL SERGEANT HANDBOOK

- Explain the responsibilities of a platoon, squad, or team leader to Soldiers. Inspire them to take on the challenge.

- Soldiers want to be challenged (both mentally and physically). Give them the opportunity to excel by creating different challenges.

Common sense techniques:

- Understand that training support packages (TSPs) are guidelines and might not always relate to current situations. Use your personal experiences, paint a picture or visual, and make training realistic.

- "Smoking" the Soldiers is not always the right answer. A firm talk, having the Soldier write an essay, taking away a privilege, even getting a parent on the phone might be more effective. Seek clear guidance from your chain of command first.

- Know when to raise your voice to make a point. Do not be someone who is always screaming, because the Soldiers will tune you out. "When in doubt, don't scream or shout!" Be direct and clear without yelling or using profanity—take the high road.

- Know the difference between being strict/enforcing standards and being abusive/demeaning. Strict drill sergeants motivate Soldiers; demeaning drill sergeants shut Soldiers down.

- Every unit has that one drill sergeant who is just one mistake away from being relieved. Do not fall for it just because he is getting away with something. It is not okay for you to do it.

- When a Soldier compromises the Army Values, use one of the Army's outstanding Soldier stories (such as SFC Smith who was awarded the Iraqi War Congressional Medal of Honor) to inspire him.

- Challenge the Soldiers with nightly assignments from their smart book or the Skill Level I Common Tasks Manual. The assignments can be based on upcoming training. Use drill sergeant time to educate and quiz the Soldiers. Demonstrate how they can put their basic knowledge to use. This technique is very effective, and the Soldiers are usually excited about learning.

- Print and put additional FMs into binders for use at the field training exercise. Have the Soldiers teach from the manuals with your supervision. It gives them a chance to get in front of their peers and show what they have learned. Many of them are very good at it—they just need a little support and confidence.

- Demonstrate everything you can—obstacle courses, first aid, land navigation, and so forth. Show Soldiers that you are a master of these tasks and can perform them yourself. Your are not just talk—you are a Soldier too!

- Talk to your Soldiers about your experiences in the Army, especially in combat. Fear of the unknown is always a demotivator.

- Maintain and account for all equipment. Check and re-check your hand receipts. Assign Soldiers to assist to build leadership/accountability.

- Conduct an inventory after all major training events, end of cycle, or when any equipment is turned in. Develop an equipment smart book for hand receipts, inventories, and statements of charges; and keep it updated.

- Keep a leader's book that includes pertinent information on all of your Soldiers (hot/cold weather injuries, medical profiles, mandatory release dates, etc.).

- Always focus on bay area security. Ensure your Soldiers know how to secure their personal items properly. Let them know they should report any missing or stolen property to you immediately.

- Be cognizant of your unit's hazardous material requirements. Ensure all hazardous material is stored properly in accordance with Army and installation policies, regulations, or guidance.

- You may encounter Soldiers going through various life changes or emotions such as homesickness, withdrawal from certain medications, anxiety, depression, anger, fear, or resistance to discipline. Seek clear guidance from your chain of command immediately on how to handle these challenges.

- Become familiar with the Community Mental Health Services (CMHS) process. Get your chain of command involved to ensure Soldiers get proper treatment or counseling.

- Deal with disciplinary issues up front. Do not let the Soldiers talk back to you—correct them on the spot! This sends a message to the other Soldiers that this behavior will not be tolerated.

- Make all instructions easy to understand. Link Soldiers who are struggling to comprehend with a buddy who can help them. Follow up with these Soldiers after the training is complete. If possible, match your stronger Soldiers with your weaker Soldiers for successful buddy teams.

- Conduct a check on learning after each training session. Make sure you ask the Soldiers specific questions. If you ask yes or no questions (such as "Do you understand?"), the Soldiers will automatically respond with "Yes, Drill Sergeant!" (whether they understand or not).

- Do not let the Soldiers give up or quit on a task or event. Push and challenge them to meet the training standard while focusing on precision. Have patience; do not sacrifice standards to time.

- Decide early with your battle buddy who will be the primary trainer and then focus on your responsibilities.

- Emphasize teaching Soldiers the fundamentals of every task. Attention to detail is important to the successful completion of every task.

- Always factor in excess time for motor movements, vehicle maintenance, dispatching, or personal hygiene when planning your training.

- Become more knowledgeable by asking questions, reading regulations, and observing more experienced drill sergeants.

- Set the example by always treating your Soldiers fairly and equally, participating in PT, and wearing the same equipment on foot marches.

- Find out the standard at the Central Issue Facility prior to turning in or picking up equipment so that you will be prepared for what the facility will be inspecting. Ensure Soldiers receive the correct size uniforms and equipment.

- Be flexible and open-minded to new ideas, approaches, and tasks.

- Be creative in helping your Soldiers achieve their goals. Possessing an optimistic and positive attitude is critical to their success. Promote a positive environment.

- Reinforce all skills learned on a regular basis.

- Help your Soldiers build their intangible skills (confidence, pride, willpower, and resilience) and overcome their fears through training and repetition (completing road marches, finishing a PT run, passing the end of cycle Army physical fitness test, or overcoming the obstacle course).

- Keep your goals, objectives, and mission on the forefront while building your mental, physical, and emotional skills.

- Focus on your duty and remember why you are here—to lead, coach, and train the Army's future leaders!

Trainee Abuse

Do not abuse trainees or condone abuse from other drill sergeants (have zero tolerance). There is plenty of written guidance established to help keep you out of trouble. Read and follow it! Know TR 350-6, *Enlisted Initial Entry Training (IET) Policies and Administration*. The regulation can answer a lot of your questions and offer clarification on your conduct in the IET environment.

Never touch Soldiers unless you are making a necessary correction for training or safety. Do not be timid when making necessary physical corrections to Soldiers. If you act like you are doing something wrong, Soldiers will think you are doing something wrong.

Remember: Your hat and badge do not give you the right to abuse Soldiers!

> **Trainee Abuse Definition**
> **(TRADOC Regulation 350-6 Glossary)**
>
> Trainee abuse is any improper or unlawful physical, verbal, or sexual act against a trainee. (This definition does not include acts involving a trainee against a trainee.)
>
> Examples might include:
>
> - Assault.
>
> - Extreme physical training not in accordance with a program of instruction.
>
> - Extreme profanity.
>
> - Sodomy and rape.
>
> - Sexual harassment.
>
> - Extortion of money.
>
> - Any personal relationship not required by the training mission.
>
> In accordance with this regulation, only a commander can determine trainee abuse has occurred.
>
> A trainee's consent to the act in no way affects this definition.

Inappropriate Relationships

Improper relationships with IET Soldiers are unacceptable. When drill sergeants violate the standards of proper relationships with IET Soldiers, they undermine the entire value system of the Army, and their misconduct reflects negatively upon the rest of the drill sergeant corps. Remember that you symbolize the epitome of what an NCO should be.

Look out for your battle buddies, especially in the first six and last six months of their drill sergeant duty. Do not get complacement at the end of your tour. Many drill sergeants become callous, insensitive, and even cruel towards Soldiers at the end of their tour. Keep the same mentality from the first day you began drill sergeant duty to the last day.

If you are not sure about a situation, ask someone. The old belief of "it's better to ask for forgiveness than to ask for permission" does not apply in the IET world.

The most important resources a drill sergeant has are other drill sergeants. It is important to look out for each other to prevent trainee abuse and improper relationships. Go to your "battle buddy," another drill sergeant, or your supervisor for support or advice before confronting Soldiers. Female drill sergeants can often detect the inappropriate actions of a female Soldier quicker than a male drill sergeant. They can spot "parade pretty" a mile away. Male drill sergeants can also often spot inappropriate conduct from male Soldiers quicker than female drill sergeants.

Be fair and consistent. Showing favoritism will bring your judgment into question, and you will lose your credibility.

Warning signs:

- Drill sergeants previously investigated for abuse are sometimes more prone to abusing Soldiers again because they feel they got away with it once and can do it again.

- Watch your fellow drill sergeants closely in the last six months of their duty, because many are under the impression that they can beat the system and not get caught. Watch for callousness—those drill sergeants who think all new Soldiers are bad. Their patience will be gone.

- Watch out for battle buddy distress signs and signals. On certain days, you and your battle buddy will be extremely tired or physically exhausted. Make a pact to look out for one another, because this is the time when the most mistakes occur. The increase in your physical deterioration may cloud your judgment.

- When you or your battle buddy start looking over your shoulder to see who is watching what you are doing, you should not be doing it. Stop and walk away before you regret your actions.

- If you see a drill sergeant start to lose control, remove him from the situation without alerting Soldiers. Tell the drill sergeant that he has a phone call or is needed somewhere else.

- When drill sergeants act in an unprofessional manner (yelling, screaming, throwing things, getting too close to the Soldiers, etc.), stop and correct them on the spot. Standing by and watching while your peer loses control is unacceptable.

- When your buddy does not heed your warnings, write the issue down and go to your senior drill sergeant or your first sergeant.

- Do not put yourself in any compromising positions. Do not ever be alone with the Soldiers (make sure they have their battle buddy with them).

- Know the following warning signs of a drill sergeant who may be acting inappropriately:

    ○ Gets offended when another drill sergeant enters his bay.

    ○ Does not want to talk to another drill sergeant about a certain incident or behaves as if he has something to hide.

    ○ Constantly talks to certain Soldiers on nonmilitary topics.

- Starts coming in late for work.
- Hangs around the bay areas for no particular reason.
- Shows favoritism or becomes too friendly toward a particular Soldier.
- Does not spend time with his Family when he is off duty.
- Drinks too much.
- Allows smiling, laughing, or giggling from Soldiers.
- Uses excessive profanity around or directly to the Soldiers.
- Jokes about sleeping with his Soldiers.
- Borrows money from or lends money to Soldiers.
- Purchases items for Soldiers.
- Over disciplines his Soldiers.

Separation Process

Prior to considering this step, you should exhaust all your other resources such as counseling, retraining, or restarting in another unit/platoon:

- Show your support for the Soldier without jeopardizing your integrity.
- Counsel every violation and keep it on file.
- Keep a notebook and write every shortcoming down. You might not be able to address it right at that moment; use it to remind you to make the correction later.
- Do sufficient and proper counseling so that a separation packet is not rejected for lack of evidence.

Your command deciding to retain the Soldier is not a lack of trust in you; talk to your chain of command about their decision. Do not interpret a decision to retain as a directive to graduate.

Track separation packets from beginning to end and know the process thoroughly (initial counseling, rehabilitative transfers, mental health procedures, etc.). It can save you and the Army a lot of time.

If Soldiers are assigned to a rehabilitation unit, visit them to help motivate them to heal and return to duty.

Chapter 2

Skills

Effective Counseling

I am a proud team member possessing the character and commitment to live the Army Values and Warrior Ethos

It is extremely important for drill sergeants to counsel their Soldiers. Counseling is a great tool that is too often overlooked in the initial entry training (IET) environment. Counseling your Soldiers for good performance will set them up for success. Many drill sergeants are quick to counsel their Soldiers when their performance is substandard, but it is important to counsel Soldiers on their outstanding performance as well. It will definitely mean a lot to the Soldier, and it will show them that the Army acknowledges Soldiers who go above and beyond the requirements.

Do not wait until the last minute to counsel your Soldiers. One of the hardest things for a new drill sergeant to do is to keep up with counseling Soldiers. Get some sample counseling forms from other drill sergeants, and begin saving the forms from every counseling that you do. That way when a situation arrives, you will be prepared.

Before you begin to counsel a Soldier, explain the counseling form to him to ensure he understands it. It may be easier if you relate the form to a report card (their performance = their grade). Remember: You cannot hold the Soldiers to a standard you have not set for them.

Counsel all Soldiers who have been placed in leadership positions so they fully understand what you expect of them. The counseling session will also give them an understanding of what to expect in the operational Army and show them where they are, where to improve, and what they are doing well.

Discuss the Soldier's strengths as well as weaknesses based on your observations and his peer evaluations. Give him guidance on how to improve his performance. A sample Drill Sergeant Peer Evaluation Form is located at Appendix B (formats may vary based on installation) and can be found in the Directorate of Basic Combat Training Center of Excellence (drill sergeant program proponent) AKO Knowledge Center at <http://www.us.army.mil/suite/portal/index.jsp>.

The Soldier Assessment Report for IET Soldiers at Appendix C is another helpful tool that assists drill sergeants in verifying enlistment eligibility and in identifying Soldier leadership and personal readiness issues having a predictable, direct, and substantial impact on IET. For use of this form, see United States Army Training and Doctrine Command (TRADOC) Regulation (TR) 350-6, *Enlisted IET Policies and Administration,* or visit the TRADOC Website at <http://www.tradoc.army.mil>.

Time and Stress Management

Maximize time during the training day to set Soldiers up for success. This may mean additional face time in the beginning of the cycle, but it will pay off in the end. The more effort you put into your Soldiers, the better the effort you will receive from them every day. Conduct precombat checks and precombat inspections prior to executing any training. Use end of cycle after-action review comments to help improve your performance.

Stress is going to be a part of your everyday life. The most important thing to remember is to always deal with stress constructively. When it comes to training, remember that stress should always be between the Soldier and their task, not between you and the Soldier. The Army has a wide variety of resources to help deal with stress. The Drill Sergeant Wellness Program is discussed in TR 350-16, *Drill Sergeant Program,* and is a great tool for you to use.

You can help your battle buddy and his Family members deal with a variety of difficult situations, from preparing for a new baby and stress management to finding help for domestic violence, child abuse, or sexual assault from various outside agencies (Army Community Service, Army Emergency Relief, chaplain, Red Cross, etc.). You can also utilize those same agencies to assist your Soldiers with Family emergencies, coping issues, or financial problems. Sit them down and find out the issue; show them that you care. Contact your command for guidance.

Read and implement all training support packages (TSPs) in line with unit guidance. When used properly, TSPs give you the information and guidance to conduct training properly; knowledge helps you teach better. The latest versions of the TSPs are available through AKO or your unit first sergeant.

Manage your time wisely. There may not be enough hours in the day for you to accomplish all of your tasks unless you prioritize and work with other drill sergeants. Focus on time management and multi-tasking. Do not spend too much time on one single thing (Soldier issues, taskings, range coordination, etc.); you may be ignoring the other ten important issues occurring at the same time. Trust your Soldier's leadership and use them during each phase of training.

Duty Platoon/Week/Day

Duty platoon is a program run by the unit first sergeant that directs a certain platoon (on a rotational basis) to provide support for the company during a certain time frame (day or week). The support includes providing personnel for any details and range support, coordinating training events, and transporting Soldiers to and from training events.

Stay flexible and adaptable—change is constant. It will be a challenging time (especially during basic rifle marksmanship), but the number one rule is not to forget anything. It is vital to keep a checklist or tracking sheet with you. Always know what is going on and the location of all your personnel.

Communication is vital—keep your command (commander, executive officer [XO], and first sergeant) informed. If you are in charge of duty week/day, make sure all training is planned and executed ahead of time. Check and then re-check to ensure that the entire duty runs smoothly, and there is sufficient reaction time to correct any issues.

Know your responsibilities and the responsibilities of your command. The XO is responsible for chow (meals, ready to eat; hot meals; dining facility meals) and the setup/coordination for ranges. You are responsible for ice sheets, training aids, any tasks, and transportation.

Stress teamwork. Ensure Soldiers police up after themselves. Use your junior leaders to check the area after training is complete.

Teach your Soldiers how to read and use the Wet Bulb Globe Temperature Index Calculator (illustrated in Chapter 3). They can assist in monitoring the temperature and possibly prevent a heat injury from occurring.

Once you direct a task, you must follow up and observe to ensure it is completed correctly. Do not be afraid to delegate tasks. Clearly explain your requirements so there are no misunderstandings.

Conduct coordination meetings with your key personnel (especially your platoon sergeants) at every opportunity. If time permits, conduct a rock drill with them (factor in travel time to and from training locations).

Brainstorm with your peers. Brainstorming can help you catch and fix problems.

Make sure you rehearse any scheduled briefings or classes you are designated to give. Appoint a demonstrator prior to the briefing or class and practice with him.

Check the Range Facility Management Support System (RFMSS) at least 72 hours prior to ensure all coordination is in place.

Range Setup

Keep your first sergeant informed. He has overall responsibility of the range. Coordinate and set up as soon as you are notified of your range duty. The more you prepare, the smoother it will go.

Use a range checklist/book or continuity book to help you remember each of your tasks. Check to ensure transportation coordination is locked in, trucks are loaded with the proper equipment, and a risk assessment has been done. Double-check the risk assessment to ensure it is correct, signed, and dated. Make sure it goes to the range with the range detail.

Make sure you know how many Soldiers are assigned to the ammunition supply point detail to load magazines, fill coolers, get targets up, and break down ammunition. Put the Soldier leader in charge once you have thought through the task.

Do a face-to-face linkup with range/tower cadre prior to your range briefing. They will provide you with the range standing operating procedures and any special instructions for that particular range (procedures may vary at each range). Go to the site 24 hours prior. Give your Soldiers proper range safety procedure briefings and instructions. If time permits, do a dry-run with all participants.

Meet the ammunition truck at the designated time and place on the range (be there 30 minutes prior). Make sure the range is serviceable and the targets are functioning properly, especially if it rained the night before.

Do not forget a staple gun and any other equipment you will be using. Check the equipment early to ensure it is functioning properly (especially fire extinguishers). Remind the charge of quarters (CQ) to prep the targets/equipment.

Basic Rifle Marksmanship

Patience is the key, especially during BRM. Maintaining a calm attitude will carry over to the Soldier and have a positive outcome on what you are trying to achieve. If you are calm, the Soldier will be calm.

Coaching, teaching, and counseling techniques:

- When the Soldiers zero their weapons, have them write their adjustments down and put them somewhere on their weapons for future reference. Help the Soldier determine which eye is dominant.

- Use different colored markers during BRM to identify the Soldiers' various grouping and zeroing shot groups.

- When Soldiers are having a difficult time at BRM, get down next to them and "coach" them. Talk calmly to them; reassure them that they can do it while keeping them focused on their mission.

- Constantly rehearse and capitalize on training time. The "dime washer" exercise is an effective technique used to help Soldiers improve during BRM.

- Make magazine changing, malfunction, and clearing drills a fun competition. All Soldiers gain the skills they need to avoid wasting valuable time during qualification while boosting their confidence levels and becoming proficient. Mastering these skills will set them up for success.

- Try not to intimidate the Soldiers while they are firing. For best results, consistently practice relaxation tips such as clearing the mind, using steady breathing, applying a smooth trigger squeeze, and forming a steady but comfortable position. Holding the weapon steady helps build their muscle strength and avoid muscle fatigue. To better evaluate the Soldiers, have them practice the firing position while lying on their physical training (PT) mat.

- To enhance self-reliance, confidence, and responsibility, have the Soldiers review and interpret their own zero cards and make weapons adjustments under supervision.

- Cognitive muscle memory and repetition are vital during this phase. Introduce BRM as soon as Soldiers arrive to get them mentally prepared.

- Use the target box exercise to help the Soldiers improve their grouping and zeroing skills.

- Always monitor the Soldiers to ensure they are following the proper procedures.

DRILL SERGEANT HANDBOOK

- Drill the basics; BRM fundamentals are vital to their success.

- Be careful not to overwhelm the Soldiers with information. They are getting a lot of instruction upfront and may not remember most of it.

- Preach safety throughout the entire BRM process: Do not point the weapon at anyone. Only point the weapon at a target you intend to shoot.

- As soon as the Soldiers get their weapons, show them how to maintain them, clean them, and carry them. Go over the "do's and don'ts."

- Constantly work on and reinforce the fundamentals of BRM (especially the shot group). Understanding the concept of how to get the bullet to hit the target is very important. Continuously drilling it and reviewing it daily will help Soldiers automatically react with the correct action. Break down the instruction properly, and review it daily. It is important that the Soldiers comprehend. Teach ballistics to Soldiers so they understand what the weapon is doing.

- Do not approach the engagement skills trainer as a waste of time. It helps you detect and fix problems before the Soldiers get to the actual range. You can see exactly what the Soldiers are doing correctly or incorrectly.

- Make sure the Soldiers' gear fits them correctly so it does not interfere with zeroing and qualifying successfully.

- During BRM, a Soldier will face many obstacles such as not knowing which eye is dominant, not remembering his site, or not using his front/rear sites correctly. It is important that the Soldiers keep the same site picture each time. Teaching them the fundamentals and continuing to drill the basic concepts will eventually pay off.

- Do not use the "tape over the eye" approach; it becomes a crutch for the Soldier and a shortcut to training for the drill sergeant. The Soldiers must be able to fight in combat using and relying on the skills and abilities they obtain through training and practice.

- Conduct grouping and zeroing using various drill sergeants from different platoons. This gives each drill sergeant within the company a chance to cross-train with other Soldiers and provides an alternate approach to learning.

- As early as possible, place the Soldiers in interceptor body armor when firing so they can start getting comfortable wearing it.

- Build a mock range in your battalion area to include various stations and areas. The mock range allows for additional training on BRM fundamentals and techniques while giving the Soldiers a realistic view of range requirements.

- Give the Soldiers immediate feedback; let them know exactly what they are doing wrong and how to correct it.

- Build the Soldiers' confidence with the weapon. They must trust themselves and conquer their fear of the weapon.

Do not get frustrated. Continue to teach, coach, and counsel (stick to the process) the Soldiers until they qualify. Let the Soldiers know you are proud of them.

Warrior Tasks and Battle Drills

Warrior tasks and battle drills are critical tasks and drills required for all Soldiers to train. Applying the warrior tasks and battle drills to your specified training ensures Soldiers are proficient in their required tasks.

Conduct training sessions with your Soldiers using Figure 2-1 below. Sometimes the Soldiers get a better understanding of their requirements from a visual aid.

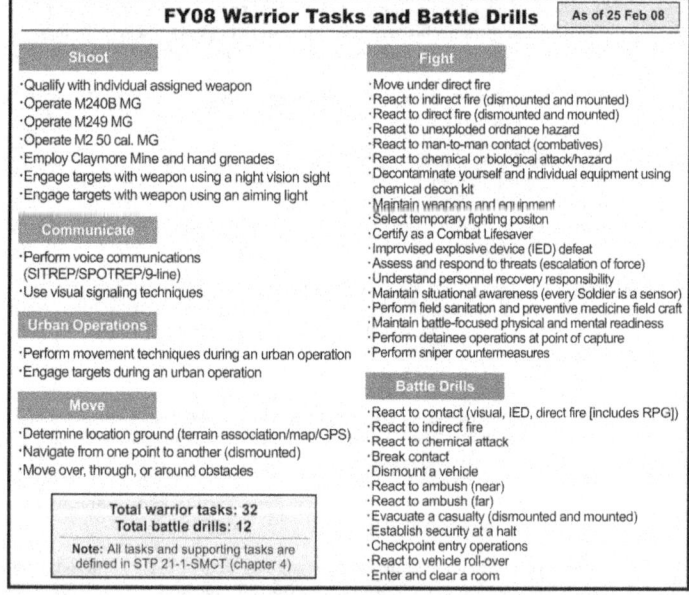

Figure 2-1

DRILL SERGEANT HANDBOOK

Charge of quarters

Be clear on your CQ duties and responsibilities. Read the unit CQ book for instructions, and do not neglect your tasks for supporting the duty platoon. Coordinate with the CQ from each company for assistance or advice (especially if you need to leave the company area). CG duties include the following:

- Stay awake! Check on your Soldiers. Ensure they are focused on the next day's training and studying with a buddy. Quiz them with training questions.

- Double-check the duty roster to ensure duties (CQ runner, fire guard, etc.) are assigned equally and fairly to each Soldier.

- Make sure you complete your accountability checks prior to performing "lights out" procedures or issuing weapons (always know the status of every Soldier).

- Complete all of your checks (especially personnel and weapons) on time and closely follow your checklist.

- Continue to monitor the detail roster to ensure CQ runners and fire guards are performing their assigned duties to standard.

- Ensure all door alarms are functioning properly to deter Soldiers from sneaking out and to prevent unauthorized entry.

- Monitor your Soldiers. If you catch someone being disruptive in the middle of the night, do not wake up everyone, but give that Soldier an impromptu class on respect—tie it into the Army values.

- Make sure your last shift of CQ runners and fire guards are in their PT uniform.

Sick call

Send Soldiers on sick call to the aid station so they do not miss an entire day of training. Make sure they have a buddy with them (normally Soldiers going on sick call will become buddy teams). Do not ever deny a Soldier's right to go on sick call.

Some Soldiers will try to hide their injuries. If you see your Soldiers limping, send them to sick call immediately. This practice could prevent an even bigger injury.

Watch for those Soldiers you suspect of malingering or pretending to be sick. Save all of their sick slips in addition to counseling them.

Tactical training

Incorporate tactical training into everything you do. Practice patrolling whenever you enter or leave the company training area. Show the Soldiers how to form a wedge, file formation, or other tactical movement. Conduct proper security procedures (choke point, secure the area, accountability, count out, various hand-and-arm signals, noise discipline, or room clearing) when arriving or departing the PT field or classroom.

Tactical training techniques:

- Always maintain proper command and control.

- Apply the crawl-walk-run method of instruction.

- Choose an area conducive to training (under a tree or shaded area, in the bleachers, or in a classroom). Have the Soldiers form up around a sandbox, and use pine cones, rocks, or sticks to simulate their roles/positions in the training.

- Take the Soldiers step by step through the entire tactical training process, but keep it simple. Remember: When training Soldiers, they will believe everything you teach and tell them. Make sure the information you put out is correct—it could save their lives one day.

- Always review what the Soldiers have learned prior to proceeding to the next step. They will eventually start to see the big picture.

- Once you have completed the sandbox demonstration, physically walk through the process with them. Continuously make corrections to ensure the Soldiers are confident and knowledgeable at each step.

- Automatic performance results from continuous practice and mitigates issues or problems. Pre-train the day before on the next day's training. Place Soldiers in different positions so they will be familiar with each role (they will not always be in the same position).

- Beginning tactical training in an open field helps Soldiers see what is required in each position. Gradually add more protective posture and obstacles (woods, opposing forces, movement to contact, etc.) as you increase the intensity of training.

- Have the Soldiers physically show you what battle drills they have learned. Do not train the same tasks over and over. Mix up the tasks; make the Soldiers think. They need to understand why they are doing the tasks as well as how to do them.

- Perform situational training exercises to standard. Do not cut corners; give the Soldiers realistic and effective training.

- Conduct after-action reviews after each training event to ensure Soldiers know what they did well, what they should sustain, and what they need to improve.

- The BCT smart card at Appendix D is another great tool for Soldiers to use as a quick reference snapshot of various tactics, techniques, and procedures.

Field training

Apply all of your tactical training to the field training exercise. When you train your Soldiers thoroughly and review/practice each tactical exercise with them, their performance shines, and they operate with a high level of confidence, precision, and professionalism.

Field training techniques:

- Make the training as fun and interesting as you possibly can—Soldiers will be more alert and motivated.

- Check your Soldiers' equipment early to allow for extra time to repair or exchange the equipment if needed.

- Review the next training day with your Soldiers; prepare them for success.

- Utilize your Junior Reserve Officer Training Corps Soldiers as much as possible. They can help the less experienced Soldiers.

- Thoroughly brief the Soldiers on the importance of the buddy team.

- Emphasize attention to detail while challenging the Soldiers.

- It is a good idea to teach the phases of tactical field training to your Soldiers. It helps them mentally prepare for the next level of challenges.

- Brief the Soldiers on the importance of taking the training seriously. Sustainment military occupational specialties (human resources specialists, financial management technicians, etc.) could easily find themselves in a war zone forced to apply these very skills.

- Review videos, slide presentations, and other training aids to help reinforce tactical field training.

Chapter 3

Abilities

Confidence

I am confident, adaptable, mentally agile, and accountable for my own actions.

The confidence to be a drill sergeant and a mentor to other Soldiers has always been held in high regard. Many seek the position as a stepping stone in their military careers. Others choose the road to Drill Sergeant School to have their own personal effect on future Soldiers. Whatever the reason, good drill sergeants must possess the following attributes:

- Superior leadership and exemplary bearing, personal appearance, and physical stamina.

- Exceptional maturity, patience, human understanding, and self-control in dealing with training new Soldiers.

- Ability to speak and instruct effectively without profanity, hazing, or degradation of human dignity.

- Professional skill, competence, and knowledge in the basics of being a Soldier.

Having confidence in yourself and a positive attitude will make you a success. Many noncommissioned officers (NCOs) start their drill sergeant duty on the wrong foot. They make it clear they do not want to be a drill sergeant because they want to stay in the operational Army and and be with their friends, or they had a bad experience with their drill sergeant. The sooner you embrace your role as a drill sergeant, the better. If not, your young Soldiers will sense a poor attitude and feed off it. They will become rebellious, start acting out, and create conflicts. They say, "If you don't care, why should I?" Do the right thing!

Physical Fitness

Physical fitness techniques:

- Focus on safety and injury prevention during all physical training (PT) sessions. Make PT safe, fun, and challenging for your Soldiers while you improve their physical fitness level.

- Do PT with your Soldiers:

    ○ Soldiers like a challenge, and they will try their best to impress you.

    ○ In order to get overall good results in physical fitness, you need to be a good role model for your Soldiers.

    ○ Set the example by training with your Soldiers. The training keeps you in shape, although you may want to work out more based on your personal PT needs. Most drill sergeants experience a reduction in physical fitness due to constantly starting PT over with new and inexperienced Soldiers.

- ○ Be motivated during PT. PT gets monotonous and repetitious if you do not put in that extra effort.

- At the end of the third week of training, the Soldiers should be familiar with how to properly conduct the standardized PT (SPT) exercises. Ensure your SPT program:

  - ○ Improves the Soldiers' physical fitness while controlling injuries.

  - ○ Progressively conditions and toughens the Soldiers while developing their self-confidence and discipline.

- Stay focused on ensuring Soldiers perform the exercises to standard, and make immediate corrections when necessary.

- Stress precision during the exercises. Help them improve the execution of the exercises (train their body through form and repetition). Give more advanced PT requirements to the more physically fit Soldiers.

- In order to achieve the desired results, ensure the PT schedule is followed. The schedule is methodically sequenced to adequately challenge all Soldiers entering the Army. Explain the guidelines for every ability group (i.e., Ability Group A, B, C, or D). Give Soldiers incentives for moving up to the higher group.

- Ensure your program emphasizes progressive conditioning of the entire body and embodies the fundamental components of strength, endurance, and mobility. Make sure your program progressively increases training intensity while controlling injuries.

- Do not invent exercises; this could cause Soldier injuries or abuse.

- Continually monitor what the Soldiers are doing. Do not just stand around giving orders. Do PT with the Soldiers.

- Use platoon, squad, and team leaders to help ensure the other Soldiers conduct the exercises correctly.

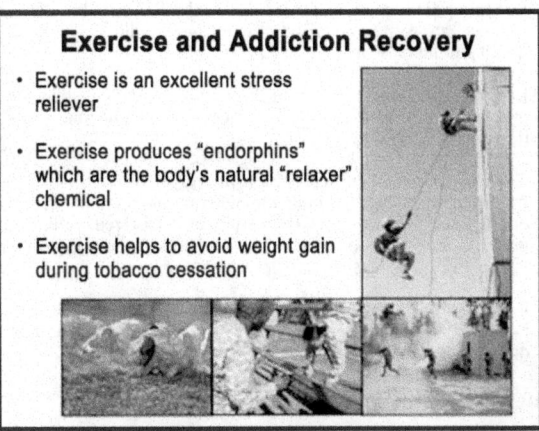

Figure 3-1

- Offer the Soldiers incentives such as earning the unit an additional PT streamer, which represents pride and discipline in the entire unit and can often result in increased Army physical fitness test scores.

- Drill sergeants should rotate into different run groups on a weekly basis. If you are always running in Group D, you are not challenging yourself. Your Soldiers will notice and lose confidence in your ability to lead from the front.

- Different training events such as combatives and pugil stick are great confidence builders.

Nutrition and Performance

Remember that your Soldiers' ability to learn and perform at their best is related to their opportunities for rest and recovery. Part of the recovery must include fueling their bodies for success.

Ask your Soldiers their performance goals and guide their food choices toward achieving these goals (these same concepts can also enhance your performance as a drill sergeant):

- **For increased energy:** Fuel with complex carbohydrates like whole wheat breads, pasta, and cereals; brown rice; and potatoes. Avoid simple sugars (sucrose and high-fructose corn syrup) that may cause a rise in energy followed by a crash. Some fat (nuts and unsaturated oils) are necessary in the diet to provide some energy and store vitamins, but limit their intake.

- **For muscles:** Fuel with lean meats like (broiled or baked) fish, turkey, or chicken; egg whites; and skim milk (ensure Soldiers include resistance training to increase lean muscle mass).

- **For stronger bones:** Fuel with skim milk, low-fat chocolate or white milk, yogurt, or calcium-fortified juices (low-fat chocolate milk is the best recovery beverage).

- **To enhance performance, repair, and growth:** Fuel with a variety of fruits and vegetables to obtain the minerals and vitamins necessary for optimal performance. Choose a variety of colors (red, yellow, orange, and green) to ensure a wide spectrum of nutrients. Choose high-fiber foods to extend the energy system and decrease hunger.

- **For hydration:** Choose nutrient-dense fluids such as skim milk, fruit juices, or vegetable juices. Vegetables and fruits also contain high levels of water. Leave the energy drinks for field training.

Dining procedures:

- Orient the Soldiers on how the dining facility and serving lines are set up to eliminate the unknown and possibly speed up the process.

- Prior to taking the Soldiers to the dining facility, task a Soldier team to recon the serving line and report back to the unit. The report should include the menu with a focus on where the high-performance foods are located.

- Ensure that during Red Phase, your Soldiers have at least ten minutes to fuel (eat). This will enable them to replace needed glycogen stores and take in needed nutrients—especially during this more stressful period of training.

- Know which Soldiers are vegetarians, have food allergies, or follow a certain religious practice. Ensure that these Soldiers have appropriate alternatives available.

- As you walk through the serving line area/dining room, look to see if your Soldiers' trays consist of a variety of foods to meet their fueling needs. You should see one-half of the tray covered with a variety of fruits and vegetables, one-quarter of the tray composed of complex carbohydrates (starches), and one-quarter of the tray in the form of lean protein.

- Encourage those Soldiers who have high metabolic rates to increase calories by consuming low-fat milk or increasing portions of calorie-nutrient dense foods like raisins, peanut butter, and mixed nuts.

Safety Awareness

Conduct a safety briefing before all training. Instill in your Soldiers that everyone is a safety officer. Soldiers should:

- Hydrate and monitor their fluid intake by using pace count beads or simple Ogden cords and placing a knot for each canteen they drink.

- Check each other's equipment.

- Keep the lines of communication open and clear and report any safety violations to the drill sergeant or another cadre member immediately.

Safety tips:

- Preventing injuries is the first priority.

- Modify training if more than one Soldier exhibits symptoms of heat injury.

- Use iced sheets for rapid cooling. When in doubt, contact medical personnel or call for an ambulance.

- Make sure the Soldiers know the importance of monitoring each other for any signs of heat or cold injury symptoms. Assess the entire group once an individual Soldier exhibits symptoms of heat injury.

- During each phase of training, check to ensure transportation coordination is locked in, trucks are loaded with the proper equipment, all personnel are accounted for, and a risk assessment has been completed. Double-check the risk assessment to ensure it is correct, signed, and dated.

- Ensure the Work/Rest and Water Consumption Table (Figure 3-2) and the Wet Bulb Globe Temperature Index Calculator (Figure 3-3) are checked and constantly monitored in order to help prevent any heat casualties. Know how to read and use the Work/Rest and Water Consumption Table and teach it to your Soldiers—it can save their lives! Use your installation card if one is available.

- Do not forget to monitor yourself and your battle buddies. The Soldiers get a lot more breaks than you do.

- Keep packs or cups of ice in your office freezer or cooler for Soldiers to use on sore muscles (it can also cut down on sick call visits).

CENTER FOR ARMY LESSONS LEARNED

Work/Rest and Water Consumption Table

Applies to average sized, heat-acclimatized Soldier wearing BDU, hot weather. (See TB MED 507 for further guidance.)

Easy Work	Moderate Work	Hard Work
• Weapon Maintenance • Walking Hard Surface at 2.5 mph, < 30 lb Load • Marksmanship Training • Drill and Ceremony • Manual of Arms	• Walking Loose Sand at 2.5 mph, No Load • Walking Hard Surface at 3.5 mph, < 40 lb Load • Calisthenics • Patrolling • Individual Movement Techniques, i.e., Low Crawl or High Crawl • Defensive Position Construction	• Walking Hard Surface at 3.5 mph, ≥ 40 lb Load • Walking Loose Sand at 2.5 mph with Load • Field Assaults

Heat Category	WBGT Index, F°	Easy Work		Moderate Work		Hard Work	
		Work/Rest (min)	Water Intake (qt/hr)	Work/Rest (min)	Water Intake (qt/hr)	Work/Rest (min)	Water Intake (qt/hr)
1	78° - 81.9°	NL	½	NL	¾	40/20 min	¾
2 (green)	82° - 84.9°	NL	½	50/10 min	¾	30/30 min	1
3 (yellow)	85° - 87.9°	NL	¾	40/20 min	¾	30/30 min	1
4 (red)	88° - 89.9°	NL	¾	30/30 min	¾	20/40 min	1
5 (black)	> 90°	50/10 min	1	20/40 min	1	10/50 min	1

- The work/rest times and fluid replacement volumes will sustain performance and hydration for at least 4 hrs of work in the specified heat category. Fluid needs can vary based on individual differences (± ¼ qt/hr) and exposure to full sun or full shade (± ¼ qt/hr).
- **NL** = no limit to work time per hr.
- **Rest** = minimal physical activity (sitting or standing) accomplished in shade if possible.
- **CAUTION: Hourly fluid intake should not exceed 1½ qts. Daily fluid intake should not exceed 12 qts.**
- If wearing body armor, add **5°F** to WBGT index in humid climates.
- If doing Easy Work and wearing NBC (MOPP 4) clothing, add **10°F** to WBGT index.
- If doing Moderate or Hard Work and wearing NBC (MOPP 4) clothing, add **20°F** to WBGT index.

Figure 3-2

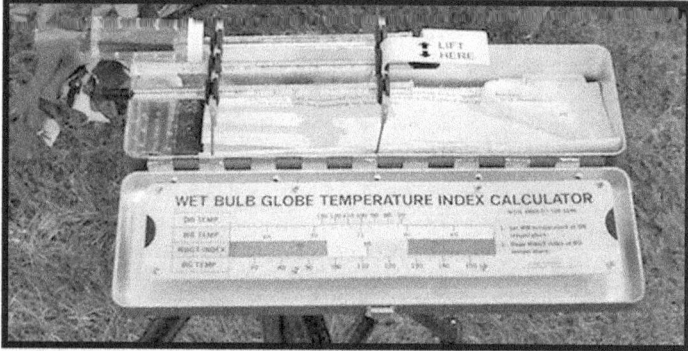

Figure 3-3

Self-Discipline

I am a self-disciplined, willing, and adaptive thinker capable of solving problems commensurate with my position and experience.

People learn in different ways, and some Soldiers may require more assistance than others. Be creative and flexible when showing Soldiers how to accomplish a task.

Never underestimate Soldiers. They are constantly watching everything you do. Always do the right thing. Make sure you are in the right place at the right time and are mentally ready for any mission regardless of the challenges. The impression that you leave on the Soldiers can have a profound impact on them for the rest of their lives.

Learn your responsibilities as a drill sergeant. Pair up with one of the more experienced drill sergeants in your unit. Make sure you read and familiarize yourself with United States Army Training and Doctrine Command (TRADOC) Regulation (TR) 350-6, *Enlisted IET Policies and Administration,* as soon as possible. Consult your chain of command for any variations within your unit. Get certified in everything required by the unit as soon as possible (mail handling class, bus, 5-ton truck, ranges, etc.). Remember that as long as you are not qualified, your "battle buddies" are pulling your slack.

Chapter 4

Attitude

Dedication

*I am physically, mentally, spiritually, and emotionally
ready to fight as a ground combatant.*

Remember to stay disciplined. Your Soldiers will have a natural tendency to emulate you whether your example is good or bad. Uphold and maintain the Army standards. You truly set the standard for your Soldiers. Make sure the standards you set are the right ones.

Your dedication to the Army and your Soldiers must be the most important item on your agenda. As you prepare them for combat, remember that your Soldiers depend on you to give them the highest quality training and guidance. Do not let them down!

Always be firm but approachable. It is easier to be hard at the beginning and then gradually tone it down (if necessary) as training progresses.

Self-Evaluation

*I am a master of critical combat skills and proficient
in basic Soldier skills in all environments.*

Drill sergeant duty is a time of constant and honest self-evaluation. Nowhere in the Army is it more necessary to train properly than during drill sergeant duty. The Soldiers placed in your charge must be trained in the basic Army tactics and standards. Your honest self-evaluation and willingness to accept positive criticism will ensure you constantly improve as a trainer, enforce standards, and be a role model to your Soldiers.

You must be receptive to new ideas and absorb information from all the other drill sergeants. After-action reviews will play an important role in identifying strengths and weaknesses in training. Do not take comments personally; correct them and drive on.

Choose thoughts over emotion to guide your behavior—that is the professional mindset of the noncommissioned officer. Show the Soldiers that you are a rock; control your actions. If you have an emotional reaction to every single Soldier action, the Soldiers will see it as a sign of weakness and think they can manipulate you.

Communication is essential. If you have a disagreement with one of your fellow drill sergeants, do not let that affect your performance. When you are not on the same page, it creates confusion, and Soldiers do not normally perform to their potential. Keeping your concerns bottled up could result in saying or doing the wrong thing, which could jeopardize your career.

Evaluate yourself. Make sure your boots are broken in; too many new drill sergeants get hot spots, shin splints, and knee and ankle pain from being on their feet too long.

Positive Leadership

The Army will train you in Drill Sergeant School on positive leadership. This is not a softer method of training—it is the Army standard. (See Major General John M. Schofield's Definition of Discipline at Appendix E.)

Positive leadership techniques:

- When problems arise, address the Soldier's inappropriate behavior or standards the Soldier did not met.

- If you are in a gender-integrated environment, make sure you know the needs of the genders as well as your limitations.

- Take ownership of your Soldiers. Treat them with respect. Show them you care.

- "This is the way we've always done it" is not a reason for training; it is an excuse for not knowing the reason. Understand all of the training you do and be prepared to explain the reason for the training. Understanding is part of positive leadership.

Avoid Burn-out

Techniques to avoid burn-out:

- Focus on teamwork, and keep communication open. Share responsibilities with your battle buddy. Delegate or trade off tasks (when possible). Give a good backbrief before and after any duties or tasks.

- Rotate taking power naps.

- Do not drink caffeine or power drinks regularly to perk up. Avoid the use of alcohol to calm down.

- Use an early or late person schedule so that everyone gets an equal amount of breaks. Balance your Family and work.

- Rotate Sunday duty with your buddies.

- Let your battle buddies leave or take a break if you do not need them.

- Master time management. Do not be a procrastinator; jump on your tasks and plan early. Get organized. Make a "to do" list; prioritize your actions; lay out your uniform, boots, and so forth the night before; and use a checklist for each phase or task.

- Seek medical attention when you need to (small problems can easily become bigger ones). Keep your shots up to date.

DRILL SERGEANT HANDBOOK

- Learn from your mistakes and do not take things personally.
- Stay motivated! Your enthusiasm will spread to your Soldiers.
- Take pride in everything you do.
- Eat and sleep when you can.

Drill Sergeant Wellness Program—Balance Family and Career

Swap out time with battle buddies or other drill sergeants to take care of personal Family matters or issues without jeopardizing the mission. This arrangement helps maintain stress levels and provides additional Family time to everyone.

Make time for your Family—they need you as much as you need them. Make sure you attend Family events. Call your Family during your personal time or scheduled breaks—they really appreciate it (especially your children).

You must have a healthy Family life. Do not bring your problems, anger, or frustrations to work and take them out on your Soldiers. Do not bring your work frustrations home and take them out on your spouse or kids.

Take advantage of cycle passes to get a stress break. During your cycle break, update your records and take care of anything you have been putting off. Do not assume cycle breaks are just time off; communicate what you want to accomplish with your chain of command. Without input from you, your commander or first sergeant will fill the schedule with what needs to be done for the unit and may not be able to make changes later. If possible, plan a Family function during your cycle break. Focus on making the time more memorable and enjoyable for everyone.

Motivation

Soldiers need to be motivated:

- Eat together as a team to build teamwork; do not sit separately from them.
- Visit with your Soldiers on Sundays to show how much you care about them (this also builds morale and motivation).
- Keep the Soldiers motivated by constantly setting the example. When Soldiers see the drill sergeant with a Master Fitness Badge, they also want one. They are motivated and encouraged to do their best at PT.
- Know what you can control and what you cannot control. Soldiers will be disciplined and motivated if they believe in the training or feel as strongly about it as you do.
- Realize that you can influence more than you can control. Leading by example is the best motivational tool; your Soldiers want to be like you.
- Make training a competition. The Soldiers will push themselves harder in an effort to beat their drill sergeant.

- Do not just challenge your Soldiers physically; create mental and teamwork leadership challenges as well. They are here to train, so make it a memorable and worthwhile experience. After you gain their respect, they will work harder for you.

- Find out what motivates your Soldiers (everyone is different and requires different motivational strategies). Playing trivia games with the Soldiers about any training they have done up to that point is a good learning and memorization technique. Try everything, but stay with what works.

- Memorization and repetition of the Soldier's Creed is great, but ensure they know what it means and how to live it.

Appendix A
References

Army Regulation (AR) 614-200, *Enlisted Assignments and Utilization Management,* June 2007

United States Army Training and Doctrine Command (TRADOC) Regulation (TR) 350-6, *Enlisted Initial Entry Training (IET) Policies and Administration,* May 2007

TR 350-16, *Drill Sergeant Program,* March 2008

TR 350-70, *Systems Approach to Training, Management, and Products,* March 1999

TRADOC Pamphlet 600-4, *IET Soldier's Handbook,* September 2007

Basic Combat Training (BCT) Smart Card

Center for Army Lessons Learned homepage, -<http://call.army.mil/>

Directorate of BCT homepage, -<http://www.bct.army.mil/>

IET Family Handbook, <https://www.us.army.mil/suite/page/352804>

TRADOC homepage, <http://www.tradoc.army.mil/>

U.S. Army homepage, <http://www.army.mil/>

U.S. Army Accessions Command homepage, <http://www.usaac.army.mil/>

U.S. Army Drill Sergeant School homepage, <http://www.jackson.army.mil/units/drill/index.html>

Appendix B
Sample Drill Sergeant Peer Evaluation Form

Figure AB-1

Appendix C
Soldier Assessment Report (IET Soldiers)

SOLDIER ASSESSMENT REPORT
(INITIAL ENTRY TRAINING SOLDIERS)
(For use of this form see TRADOC Reg 350-6; the proponent agency is USAAC and DBCT)

PRIVACY ACT NOTICE
DATA REQUIRED BY THE PRIVACY ACT OF 1974

AUTHORITY: 5 USC §301, Departmental Regulations; 10 USC §3013, Secretary of the Army.
PRINCIPAL PURPOSE: To assist leaders in verifying enlistment eligibility and in identifying Soldier leadership and personal readiness issues having a predictable, direct, and substantial impact on initial entry training.
ROUTINE USES: For enlistment and training purposes IAW AR 601-210 and TR 350-6.
DISCLOSURE: Disclosure is voluntary.

SOLDIER NAME AND RANK (*LAST, FIRST, MI*):
LAST 4 SSN:

BCT DS: _____	AIT PSG: _____
UNIT: _____	UNIT: _____
START DATE: _____	START DATE: _____
END DATE: _____	END DATE: _____
PHONE: _____	PHONE: _____
SIGNATURE: _____	SIGNATURE: _____

SECTION I - PERSONAL DATA
Note all factors, <u>good</u> and <u>bad</u>, that could impact Soldier's training.
Include: civilian education/training, language skills or barriers, Family issues, financial concerns, weight control issues, fitness level, and injuries sustained during training.

BCT:	AIT:

SECTION II - TRAINING OUTCOMES

Write 1, 2, 3, or "NA" for each training outcome:
1 = Below average / Needs much improvement
2 = Average / Satisfactory
3 = Above average / Excellent

NA: Training situation (e.g., inadequate time) did not allow IET Soldier to display quality often enough to rate.

	BCT	AIT
Outcome: Is a proud team member possessing the character and commitment to live the Army Values and Warrior Ethos		
Outcome: Is confident, adaptable, mentally flexible, and accountable for own actions.		
Outcome: Is physically, mentally, spiritually, and emotionally ready to fight as a ground combatant.		
Outcome: Is a master of critical combat skills and proficient in basic Soldier skills in all environments.		
Outcome: Is self-disciplined, willing, and an adaptive thinker, capable of solving problems commensurate with position and experience.		

Figure AC-1 (front)

CENTER FOR ARMY LESSONS LEARNED

SECTION III - LEADERSHIP and LEADERSHIP ATTRIBUTES
List all assigned leadership positions held by IET Soldier, indicate why position was assigned, and evaluate performance.
If no leadership positions were assigned to Soldier, indicate why, and summarize observed leadership behaviors and/or potential.

SECTION IV - OVERALL SOLDIER ASSESSMENT
Mark an "X" to indicate IET Soldier's standing in the Platoon.
If Soldier is ranked in the TOP or BOTTOM 10%, explain why.
If Soldier is ranked in the middle, note *at least* two <u>sustain</u> areas and two <u>improve</u> areas.

BCT:	AIT:
___ Top 10% ___ Middle ___ Bottom 10%	___ Top 10% ___ Middle ___ Bottom 10%

SECTION V - COMMENTS
Use this section as a continuation to provide additional information from all other parts of this form. Add any additional relevant comments and information about this Soldier not previously recorded on this form.

BCT SOLDIER	AIT SOLDIER
I AGREE (____) / DISAGREE (____) with the information contained herein. If disagree, indicate why in Comments Section above. Signature: _____ Print Name: _____ Date: _____	I AGREE (____) / DISAGREE (____) with the information contained herein. If disagree, indicate why in Comments Section above. Signature: _____ Print Name: _____ Date: _____

*Only identify relevent medical/dental issues that are already documented in the Soldier's Health Record. A Soldier's protected health information should only be recorded here when the Soldier consents or when it was provided by the Health Record custodian based upon an official need to know IAW AR 40-66. Medical Record Adminstation and Health Care Documentation. To ensure accuracy, personal medical, ental financial, or family information will be collected directly from the individual if possible. Such disclosure by the Soldier will be voluntary.

Figure AC-1 (back)

DRILL SERGEANT HANDBOOK

Appendix D
Basic Combat Training Smart Card

The basic combat training smart card is a great tool designed to give Soldiers a quick reference snapshot of various tactics, techniques, and procedures to use in combat. The smart card will be helpful during the final field training exercise and as concurrent training.

The front of the card includes:

- 9-line medical evacuation request
- Improvised explosive device (IED) 9-line size, activity, location, time (SALT) report
- Tactical combat casualty care report
- Vehicle search procedures
- Escalation of force procedures
- Detainee operations
- Law of War (Army Regulation 350-1, *Army Training and Leader Development*)

Figure AD-1

CENTER FOR ARMY LESSONS LEARNED

The back of the basic combat training smart card includes:

- Keys to defeating IED threats
- Identifying IEDs
- 5 and 25 meter scan
- Unexploded ordnance and IEDs
- Weapon status
- Weapon clearing
- Heat/cold safety
- Instructions on what to do when the media calls

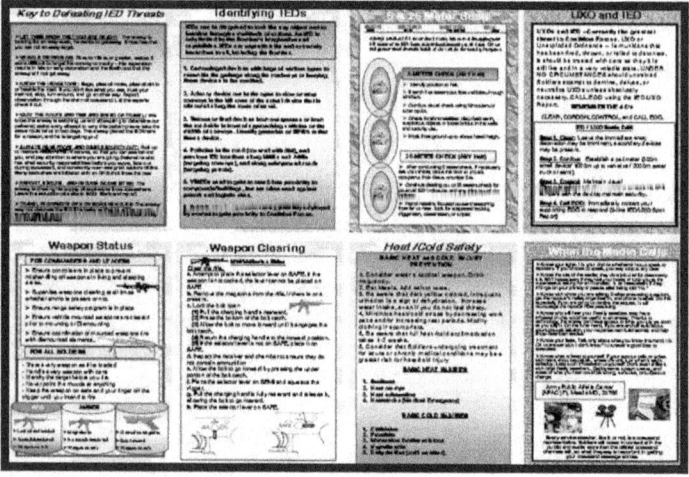

Figure AD-2

Contact your local Training Support Center to obtain the basic combat training smart card.

Appendix E

Major General John M. Schofield's Definition of Discipline

The discipline which makes the Soldiers of a free country reliable in battle is not to be gained by harsh or tyrannical treatment. On the contrary, such treatment is far more likely to destroy than to make an Army. It is possible to impart instruction and to give commands in such a manner and such a tone of voice to inspire in the Soldier no feeling but an intense desire to obey, while the opposite manner and tone of voice cannot fail to excite strong resentment and a desire to disobey. The one mode or the other of dealing with subordinates springs from a corresponding spirit in the breast of the commander. He who feels the respect which is due to others cannot fail to inspire in them regard for himself, while he who feels, and hence manifests, disrespect toward others, especially his inferiors, cannot fail to inspire hatred against himself.

DRILL SERGEANT HANDBOOK

PROVIDE US YOUR INPUT

To help you access information quickly and efficiently, Center for Army Lessons Learned (CALL) posts all publications, along with numerous other useful products, on the CALL Web site. The CALL Web site is restricted to U.S. government and allied personnel.

PROVIDE FEEDBACK OR REQUEST INFORMATION

<http://call.army.mil>

If you have any comments, suggestions, or requests for information (RFIs), use the following links on the CALL home page: "Request for Information or a CALL Product" or "Give Us Your Feedback."

PROVIDE TACTICS, TECHNIQUES, AND PROCEDURES (TTP) OR SUBMIT AN AFTER-ACTION REVIEW (AAR)

If your unit has identified lessons learned or TTP or would like to submit an AAR, please contact CALL using the following information:

Telephone: DSN 552-9569/9533; Commercial 913-684-9569/9533

Fax: DSN 552-4387; Commercial 913-684-4387

NIPR Email address: call.rfimanager@conus.army.mil

SIPR Email address: call.rfiagent@conus.army.smil.mil

Mailing Address: Center for Army Lessons Learned, ATTN: OCC, 10 Meade Ave., Bldg 50, Fort Leavenworth, KS 66027-1350.

TO REQUEST COPIES OF THIS PUBLICATION

If you would like copies of this publication, please submit your request at: <http://call.army.mil>. Use the "Request for Information or a CALL Product" link. Please fill in all the information, including your unit name and official military address. Please include building number and street for military posts.

CENTER FOR ARMY LESSONS LEARNED

PRODUCTS AVAILABLE "ONLINE"

CENTER FOR ARMY LESSONS LEARNED (CALL)

Access and download information from CALL's Web site. CALL also offers Web-based access to the CALL Archives. The CALL home page address is:

<http://call.army.mil>

CALL produces the following publications on a variety of subjects:

- **Combat Training Center Bulletins, Newsletters, and Trends**
- **Special Editions**
- *News From the Front*
- **Training Techniques**
- **Handbooks**
- **Initial Impressions Reports**

You may request these publications by using the "Request for Information or a CALL Product" link on the CALL home page.

COMBINED ARMS CENTER (CAC)
Additional Publications and Resources

The CAC home page address is:

<http://www.leavenworth.army.mil>

Battle Command Knowledge System (BCKS)

BCKS supports the online generation, application, management, and exploitation of Army knowledge to foster collaboration among Soldiers and units in order to share expertise and experience, facilitate leader development and intuitive decision making, and support the development of organizations and teams. Find BCKS at <http://usacac.army.mil/CAC/bcks/index.asp>.

Center for Army Leadership (CAL)

CAL plans and programs leadership instruction, doctrine, and research. CAL integrates and synchronizes the Professional Military Education Systems and Civilian Education System. Find CAL products at <http://usacac.army.mil/CAC/CAL/index.asp>.

Combat Studies Institute (CSI)

CSI is a military history "think tank" that produces timely and relevant military history and contemporary operational history. Find CSI products at <http://usacac.army.mil/CAC/csi/RandP/CSIpubs.asp>.

DRILL SERGEANT HANDBOOK

Combined Arms Center-Training: The Road to Deployment

This site provides brigade combat teams, divisions, and support brigades the latest road to deployment information. This site also includes U.S. Forces Command's latest training guidance and most current Battle Command Training Program Counterinsurgency Seminars. Find The Road to Deployment at <http://rtd.leavenworth.army.smil.mil>.

Combined Arms Doctrine Directorate (CADD)

CADD develops, writes, and updates Army doctrine at the corps and division level. Find the doctrinal publications at either the Army Publishing Directorate (APD) <http://www.usapa.army.mil> or the Reimer Digital Library <http://www.adtdl.army.mil>.

Foreign Military Studies Office (FMSO)

FMSO is a research and analysis center on Fort Leavenworth under the TRADOC G-2. FMSO manages and conducts analytical programs focused on emerging and asymmetric threats, regional military and security developments, and other issues that define evolving operational environments around the world. Find FMSO products at <http://fmso.leavenworth.army.mil/recent.htm> or <http://fmso.leavenworth.army.mil/products.htm>.

Military Review (MR)

MR is a refereed journal that provides a forum for original thought and debate on the art and science of land warfare and other issues of current interest to the U.S. Army and the Department of Defense. Find MR at <http://usacac.leavenworth.army.mil/CAC/milreview>.

TRADOC Intelligence Support Activity (TRISA)

TRISA is a field agency of the TRADOC G2 and a tenant organization on Fort Leavenworth. TRISA is responsible for the development of intelligence products to support the policy-making, training, combat development, models, and simulations arenas. Find TRISA Threats at <https://dcsint-threats.leavenworth.army.mil/default.aspx> (requires AKO password and ID).

United States Army Information Operations Proponent (USAIOP)

USAIOP is responsible for developing and documenting all IO requirements for doctrine, organization, training, materiel, leadership and education, personnel, and facilities; managing the eight personnel lifecycles for officers in the IO functional area; and coordinating and teaching the qualification course for information operations officers. Find USAIOP at <http://usacac.army.mil/CAC/usaiop.asp>.

U.S. Army and Marine Corps Counterinsurgency (COIN) Center

The U.S. Army and Marine Corps COIN Center acts as an advocate and integrator for COIN programs throughout the combined, joint, and interagency arena. Find the U.S. Army/U.S. Marine Corps COIN Center at: <http://usacac.army.mil/cac2/coin/index.asp>.

Support CAC in the exchange of information by telling us about your successes so they may be shared and become Army successes.

www.ingramcontent.com/pod-product-compliance
Lightning Source LLC
Chambersburg PA
CBHW050243230526
45470CB00005B/2082